Hans-Ulrich Lehmann

Design Classics

The Rings by C.-F. Dau

Verlag form

"The original will for art...is searching for pure abstraction as the only possibility to rest within the confusion and vagueness of the world view and, with an instinctive necessity, creates geometric abstraction from itself. It is the perfected expression, and the only one tangible for man, of the emancipation from everything arbitrary and transient in the world view."[1] This passage was written by Wilhelm Worringer in his dissertation *Abstraktion und Einfühlung* (Abstraction and Empathy) first published in 1908. It received a lot of attention and has been reedited repeatedly since then. And Carl-Friedrich Dau, in one of his early catalogs, outlined his concern with the following words: "Our eyes are overloaded with stimuli every day. There are very few things in which calmness is inherent that are good for the eye, where aesthetic stimulus is paired with simplicity and clarity. Creating such objects, placing them with jewelry and enforcing them in this field is the main concern of my work."[2]

The juxtaposition of these two statements does not imply that Dau's work prototypically proves Worringer's insights; this would exaggerate the intentions of any designer and minimize the highly generalized insights of Worringer. But the passages, without question, clearly describe the historic and art historic context in which Dau's work and intentions find themselves.

Preceeding page: Carl-Friedrich Dau.

Carl-Friedrich Dau was born in 1942 and is anything but a traditional goldsmith; and yet, he is among the few jewelry designers who have placed tremendous accents within the German jewelry industry within a short period of time. He lives and works in Berlin.

A Career that's Anything but Straight-Forward

Our first talk about his life and his work took place in 1993. Dau began with the statement that he really was trained as a sailor and, indeed, worked for the trade marine for four years. Back then, when he went to sea, he was 17, and the profession of the sailor seemed to him an – certainly romantic – alternative to the so-called normal life from which he wanted to escape, at least temporarily. He finished high school and took his final exam only after his seafaring experience, and then, at age 23, he apprenticed as a goldsmith with the master C. Scriba in Giessen, who was barely his senior. Dau's desire to become a goldsmith, however, developed in a rather arbitrary manner. In retrospect, he says, he could easily have become a cabinetmaker or potter. His goal was to think of things and then to produce them. Hands and mind should be occupied in a balanced way. He wanted to be idealistically and practically active and creative – an optimal combination that has meanwhile become rare. The excellent craftsmanship training he received from Scriba was followed by two years as a journeyman in Bremen with a jewelry goldsmith and, afterwards, by his first studies at the Staatliche Zeichenakademie in Hanau with H. U. Bullermann. In Hanau, design theory was taught in addition to several other subjects. The students were urged to do playful theoretical experiments with paper, scissors and pencil. Looking back, Dau remembers that these kinds of exercises did not satisfy him at all, considering the practical experience he had already acquired, and he approached them only with great resistance. Instead, he was fascinated with the movement possibilities of jewelry that he saw at the workbench, and he was literally infatuated with 'click-clack' connections and locks. And so, his masterpiece turned out to be a bracelet made of white gold with blue gold surfaces, which could be unfolded; it may have been absolutely unwearable, but he had integrated all of the technical finesse that could possibly be added. One year later, he melted it down.

From Goldsmith to Designer

In 1972, right after taking his master's exam, Dau went to Berlin in order to continue his studies within the framework of a scholarship for talented people in the trades at Berlin

University for Fine Arts. Meanwhile, he had turned 29 and had only needed five and a half years from apprenticeship to master's exam; due to his outstanding achievements, he received a scholarship that enabled him to continue his training. These studies brought about an essential change with respect to his ambitions and to his way of working: the hypertrophy of the goldsmith techniques developed in Hanau was transformed into a constructive understanding of technology. Dau was still very good in terms of craftsmanship, but suddenly it was no longer important to him to produce small technical wonders that were as spectacular as possible; his work became increasingly simplified. He had thus attained a position that would, as a general tendency, determine his future development as a designer, and to this day he has never questioned it.

Parallel to his course of studies, Dau established his first workshop in Berlin, settled himself and his family in the still-divided city and became a teacher at the training school for goldsmiths. Finally, between 1976 and 1978, he completed a third course of studies; this time, however, in the field of industrial design. Dau was 34 years old, and he occupied himself with aesthetic criteria during these additional studies. He wanted to know why we perceive an object as beautiful but can label another comparable one as even more beautiful. His dissertation was devoted to the aesthetic problems concerning the enlargement and reduction of already determined forms, and it led the designer to essential insights.

Up to this point, the endeavors in the workshop accompanied his theoretical work like a second voice. In 1980, an important decision was made that resulted in a change of priorities and led him back to his earlier goal. Dau had become a bit tired of theory, and when the occasion arose to establish a larger workshop in the basement of a villa in Zehlendorf, where the Daus lived together with two other couples they were friends with, he gave up his job as teacher and started his own business. His goal: to resolve the emphasis on theory during the past years in practice and to see whether his theoretical insights could be realized in practice. With this decision, Dau still had not really returned to jewelry in a true sense, but the foundation for today's reality had been established in many respects.

Stations – The Wall Objects

The development of today's workshop took place in three stages that can be determined in local as well as conceptional terms. During the first years of his freelance work, C.-F. Dau worked by himself and concentrated on larger objects which he assembled mainly from metal tubes and, preferably, matte glass. This work, relatively freed from utilitarian purposes, were not alone in the art scene of those days. At the same time, Friedrich Knupper realized his wall objects that can definitely be compared with Dau's early work in terms of concern and content.[3] But other artists, too, among them several goldsmiths, tried at that time to break with the seemingly narrow limitations of functionalism. At the very least, they wanted to broaden them, and it wasn't rare that goldsmiths in particular discovered the larger formats through which they were able to gain freedom

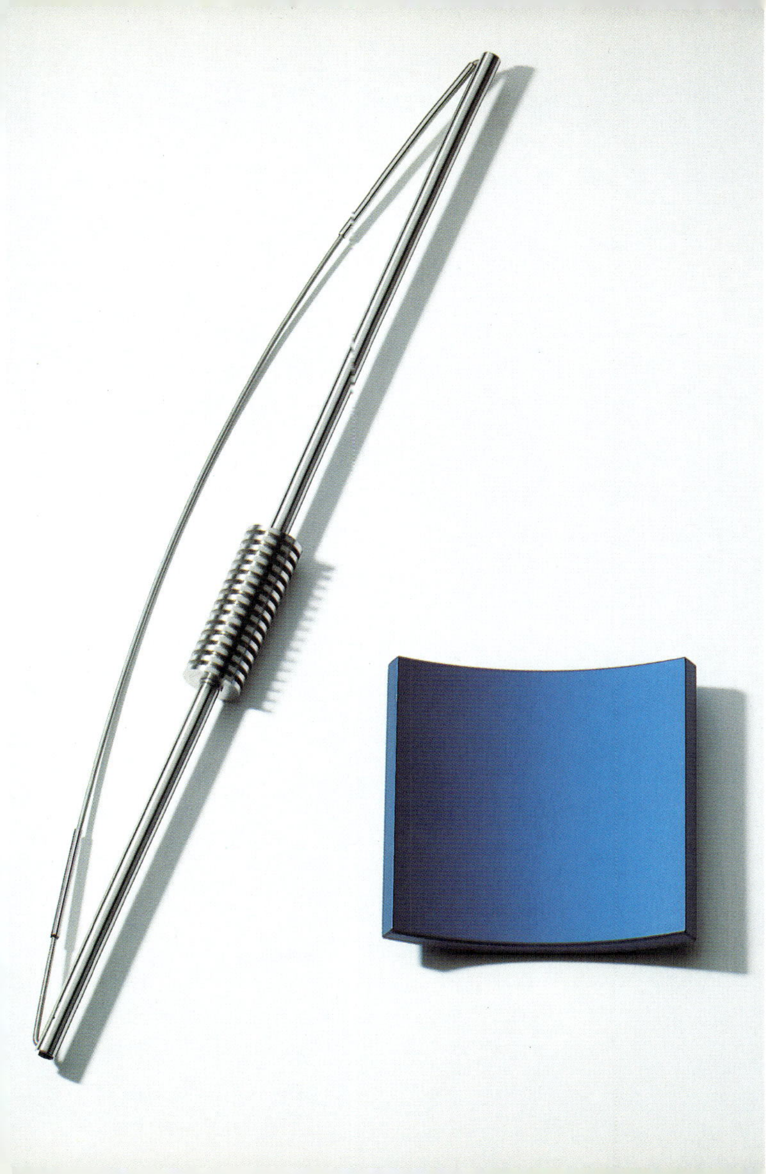

Left: Four brooches made of metal frames and differently stretched rubber bands.

Right: A bowed stickpin and a flat, colored, galvanized brooch. The flat brooch is available in three colors. The arched pin is offered in two sizes and about 40 colorful formal versions. Aluminum, high-grade steel.

for their work. The objects Dau created during this time anticipate much of what would characterize his later jewelry, but they were lacking in explicitness and functionality.

C.-F. Dau is a rather factual person, and he has a strong tendency towards constructive solutions. It would have been strange if he could have lived without a concrete purpose for a longer period of time, but the objects remained only an episode in his career despite the positive criticism they received. When he had brought them to a certain maturity, when he had learned what could be learned from work of this kind, he abandoned this field of creativity in 1983 following an exhibition in which he presented the fruits of his labor of the past years. The break with object art happened, in a sense, overnight, and from that time on he devoted himself and his work exclusively to jewelry. Functionalism as a measure for constructive activity had caught up with him; to be more precise: he had rediscovered it.

The Early Serial Jewelry

From that point onward, the development of his work took place in an extremely consequential and straight forward manner. In the beginning of creating jewelry, there was, of course, the goldsmith's enthusiasm for handcrafted creative work and the joy of seeing clear things come into being. But the task was also to determine positions with respect to design and technique and to make very concrete decisions that were important for the planned jewelry. This meant that the goldsmith standards, which had, up to that

Wall object made of steel (1m x 1m).

point, not really been reflected upon, needed to literally be lifted up. It began with the treatment of surfaces, the classic connection techniques of the goldsmiths needed to be examined with respect to their usability, and all attempts and ideas were determined by the desire to develop forms that were "as simple as possible, that are known and accessible to the observer" (Dau) and thus, to make them credible as jewelry. With this desire, the basic tenor of the seventies was taken up again, and C.-F. Dau is absolutely one of the designers who, irrespective of any fashion trends and rather untouched by special preferences of potential customers, revolve around their once-found theme every year. In an older interview, Dau once stated: "If I feel that something is beautiful and right, then I think that I am not alone with this feeling. I only have to search others who feel the same." This conviction has become a maxim for his actions; the consequential manner with which he has followed it is still remarkable to this day.

During his freelance work, C.-F. Dau never really worked on unique specimens. His ambition had always been oriented towards serial production, even though the concept of the series was meanwhile modified and specified. The first series that were developed at the beginning of the eighties were graphically dominated. They clearly echo object art and often look like cultivated and minimized wall objects. The closeness to the exercises on design theory that was endured with so much reserve then cannot be denied either, and the small objects often were actual brooches. The preferred forms were the circle, triangle and square, and

Three hanging rod pendants on stainless steel filament. Gold, silver, high-grade steel.

The new workshop and studio house, planned and realized together with the architect Jan Bremermann in a recently developed industrial area in Zehlendorf.

they were offered in different material combinations and with a wide variety of graphical structures. Despite their dependency on that era and the still perceivable textbook styling, the work on this collection did, however, open up two essential insights: the serial ambitions became concrete serial experiences, and Dau began to work with purchased parts that were prefabricated at other companies according to his drawings – at first, still with restraint and in combination with custom-made elements. But the tendency that would in the future characterize his way of working and representation became clear.

The concept worked out, the jewelry was accepted and, with the expansion of the collection, the production limitations of the one-man-business was soon reached. Qualified goldsmiths, some with additional training, could be employed, and around 1990 four employees were producing the jewelry developed by C.-F. Dau.

In 1992 – Berlin had meanwhile been reunited – the possibility arose to move part of the workshop to nearby Stahnsdorf. The assembly of the jewelry remained in Zehlendorf, but the technical processes were dealt with mainly in Stahnsdorf. Thus, space was created in the home workshop and another employee could be hired, but additional logistical necessities had emerged. Prefabricated parts had to be brought to Stahnsdorf, and other worked and refined parts had to be moved from Stahnsdorf back to Zehlendorf for assembly. Organizational expenditures increased, and a plan developed to create a single facility for the entire production. In the recently developed Zehlendorf industrial

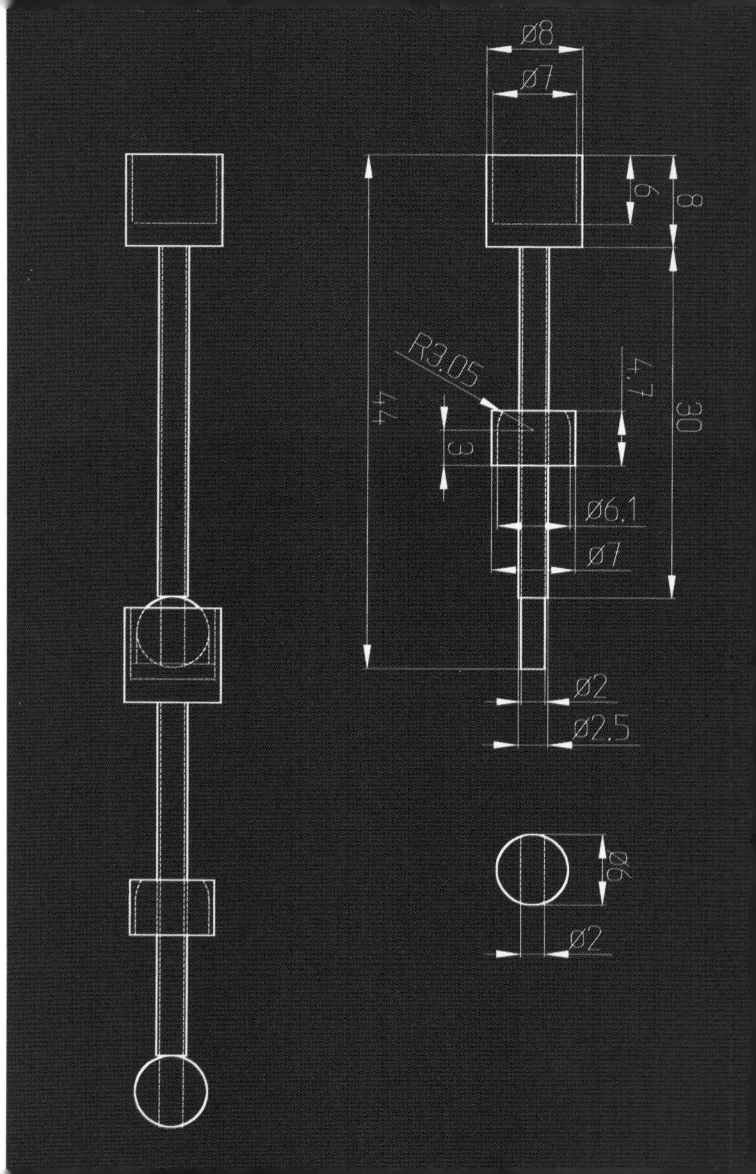

Left: Technical drawings of the ball and socket joint necklace.

Right: A necklace made of gold and stainless steel elements connected by ball and socket joints. The necklace group comprises about 15 different basic models that can be produced in numerous colored versions. The necklace shown here can also be assembled in a reverse color combination. Gold, stainless steel.

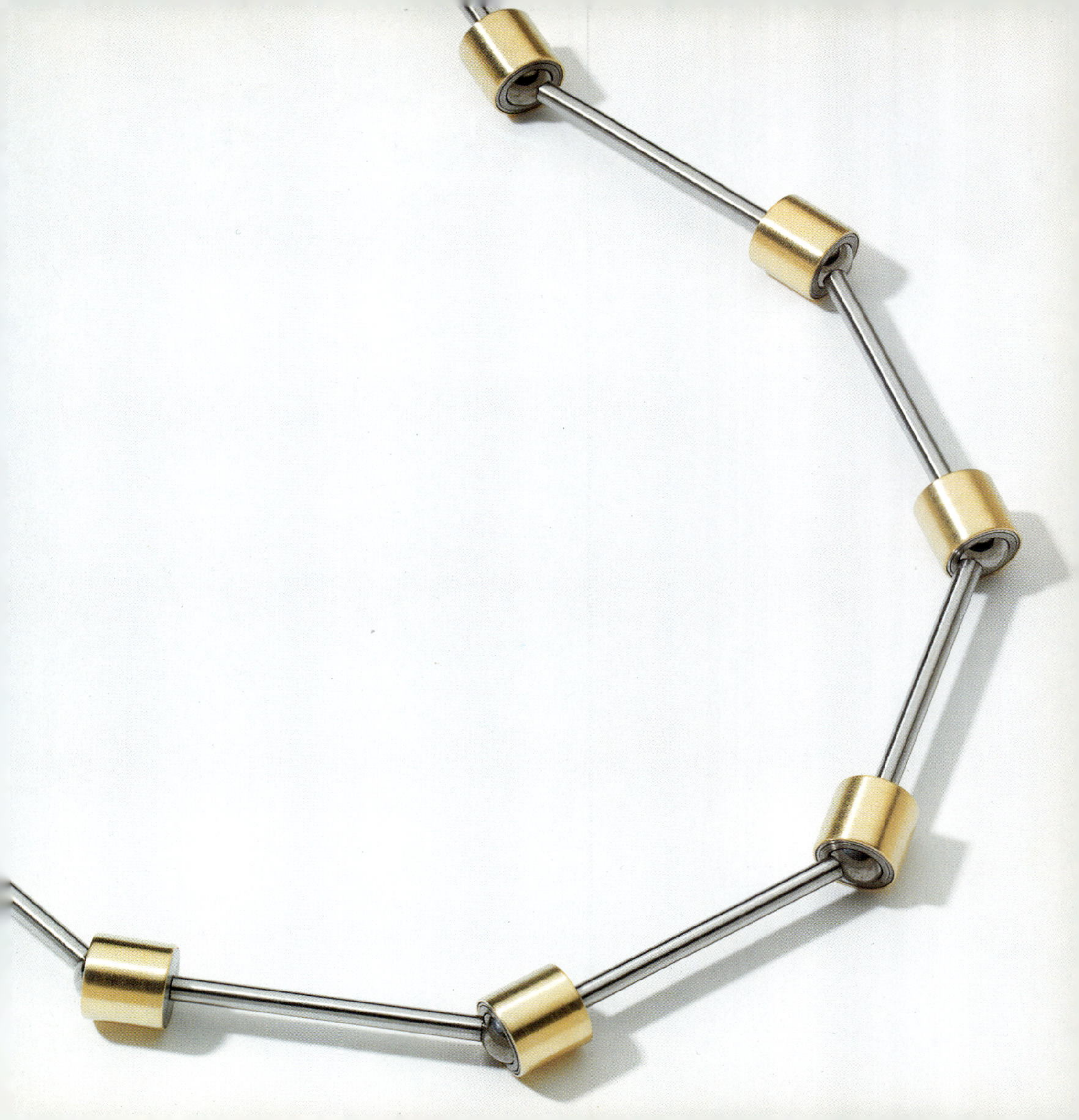

area, a property was found, and in 1994, together with the young architect Jan Bremermann, plans for the workshop and studio building at Hohentwielsteig began. The planning phase lasted twelve months. Construction was started in 1995, and by mid-1996 the generously planned and carefully realized new domicile could be furnished and moved into. Today, C.-F. Dau is working there with eight employees.

The Jewelry Since the Mid-Eighties

The styling determining today's collection developed around the mid-eighties. The early, flat, graphically dominated solutions were increasingly replaced by figurations. This development becomes especially clear when comparing the brooch jewelry. Whereas brooches consisting of framed metal surfaces or geometric wire frames dominated the first half of the eighties, the second half was dominated mainly by cylindrical metal rods. The brooches became pins and the geometric surfaces were replaced with stereometric figures or figurative forms. This development is not at all limited to the brooches. It can also be found in numerous areas of the collection – for example, in the case of the rod pendants but also the necklaces and rings. Parallel to this styling development, the collection was continuously realized using purchased prefabricated parts, and analogous to this, the flat surfaces were increasingly replaced by rotation-symetrical volumes. The polychrome coloration of the models was intensified. The round rods that dominated the picture at that time often looked rhythmically banded, and the palette was enriched by another shade of gray through the introduction of titanium.

The fundamental design theme of C.-F. Dau developed during this time is relatively easy to describe, whereas the variety of the possible readings and that of the actual solutions became larger and can be outlined only with difficulty. Dau concentrates himself and his work on few stereometric basic forms, which he orders in different metals, dimensions and several pure metal colors and juxtaposes with one another in a piece according to precise stipulations. He, together with his team, still takes care of the development from the first prototype to the actual serial production, meaning its unproblematic reproducibility. But after that, he leaves the production to those destined for it, meaning that he has individual parts produced in various companies that are assembled in his own workshop by qualified employees. This way of working presumes that the parts are clearly determined and precisely measured and described in their smallest details with respect to their position and function in the piece. Given the small size of the objects, the permissible tolerances are small, and they are especially small in the case of connections and functional parts.

The consequence of this way of working is that the expression of the final product doesn't come about during the assembly but that all special features of each pattern need to be so clearly predetermined during the model phase that the ensuing pieces can be considered valid reproductions of the original model. Strict observance of

Stainless steel necklace and brace-
let of titanium and silver.

Steel bracelets with gold central element that is movable.

this principle and the quality of its realization are essential aspects that make Dau's jewelry so special.

'Serial Art'

At the occasion of the Hannover Art Exhibition, Ernst Kállai published an article titled 'Serial Art' in 1930 in issue 11 of the magazine *Das Neue Frankfurt.* The text refers to the paintings by Carl Buchheister shown in Hanover, which can be purchased as copies personally made by the artist. Without the intention of applying to the term 'Art' to Dau's work, Kállai's remarks poignantly describe the work method and basic attitude towards design:

'All these artists [referring to Carl Buchheister, Piet Mondrian, El Lissitzky, Moholy-Nagy, Vordemberge-Gildewart, Otto Nebel and Willi Baumeister] reject the term artwork as connoting a unique occurrence determined by unpredictable and, hence, unrepeatable emotions and movements by the soul and hand. Their compositions can be mathematically measured to the last and finest detail, and they can be totally understood in their overall structure. They don't want to make anything a secret or mystery. They're playing with an open hand. They're playing: meaning that, in the end, even in these strictly constructive compositions of surfaces and space that have the reputation of being intellectualized, there remains the vibration of a purely emotional touch and a balance of elements. It's only that these elements are so clear and simple as though the emotional waves that carry them had passed through the X-ray light of reason before they appeared in the world. Hence, the exact plan of such artwork and, hence, the possibility to produce them in series in an arbitrary quantity. Without a doubt, this is a very contemporary possibility given the social and spiritual structure of our society.' [4]

Dau, too, doesn't want to make secretive things; his intention is quite the opposite. In numerous contexts he has stressed that what determines his work is the idea that carries it and that this idea has to be made tangible and transported as clearly as possible. His concern is not the emotional surface structures, not the curving of a surface determined by the crafted process, not the possibly picturesque togetherness of undertones that can be characterized only with difficulty, and what he doesn't tolerate at all are arbitrary effects that might occur during the work process. His ideas are objectified as thoroughly as possible from the first design idea to the final product and are elaborated with respect to the future production. With Kállai's words: 'the emotional waves' that carry the original plan are sent through 'the X-raying light of reason before they appear in the world.' In this way they are arranged, filtered and freed from the traces of craftsmanship. What remains is the final piece that, due to the remoteness to any possible presence of craftsmanship and emotional arbitrariness, can be reproduced in unlimited numbers. What's fascinating about this kind of jewelry is the contradiction between the almost mathematical, impersonalized regularity of the details and the extraordinarily personal appearance of the finished piece. Given the 'purely emotional touch and balance of the

The first ring in the collection of C.-F. Dau. A simple, C-shaped, 'coat of arms' band is welded onto a cylindrical rod. The rods can be produced in a variety of forms. Gold, silver, stainless steel.

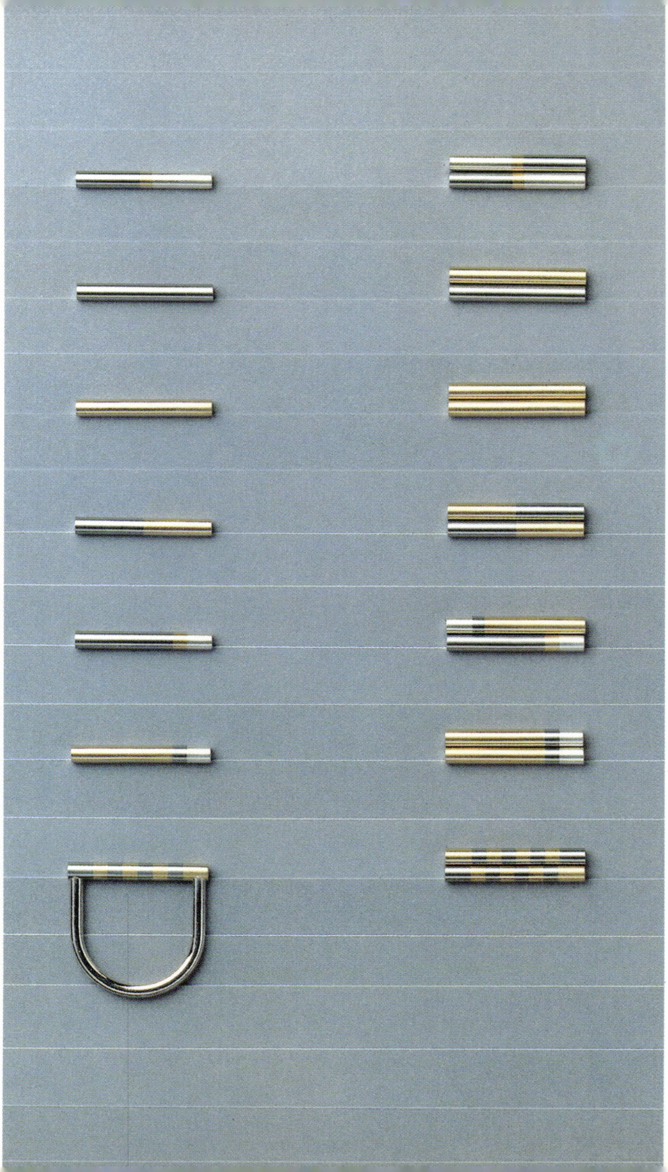

A square ring formed from four round rods. The elements of the ring are connected through press fits. This model was also admitted to the new platinum-gold collection; it is produced only in gold or only in platinum.

elements,' which, of course, is nothing if not individual, surprisingly more can come into being than the sum of the details would have us expect. Obviously, it's this difference that is responsible for the subjective and individual specialty of Dau's work.

The Rings by C.-F. Dau

In its conception and look, the ring collection by C.-F. Dau is absolutely in line with the other jewelry from his workshop, and like all other pieces in the collection it is consequentially designed for serial reproduction. The effect of most models is cool and factual, almost distant. The rings are produced or assembled from steel and silver, titanium and gold and, most recently, platinum. Their appearance mostly results from the strictly maintained band character, the original form of the ring. An aspect inherent to the band character – the stripes – causes the astonishing variation of appearances. Usually, it is created through variable metal shades and rather graphical caesuras, sometimes through discrete sculptural accents. The rings look strict and provide the appearance of their wearers with appropriate accents. The entire group of rings includes, to date, 18 basic models and a large, almost impossible to determine wealth of variety; nearly 70 possible combinations can be named. Although the band rings dominate in their numbers and variability, the ring collection is, due not least to the different eras of the creation of the individual models, not at all homogeneous. This requires a determination of the differences within the group in order to appropriately honor their uniqueness in terms of construction and design.

C.-F. Dau hasn't always designed and produced rings. The oldest model was created around the mid-eighties and is clearly related to a group of stickpins from the same time. Today, the ring is still part of the collection and is offered in monochrome as well as in a considerable number of colored variations. Constructively, it usually consists of a C-shaped round-rod band to which a simple crossbar is welded, creating a form reminiscent of a coat of arms. The solution is as self-understood as it is beautiful. When the ring is worn, it creates the impression that the crossbars, conforming to the stickpins, is simply lying on top of the finger. A ring assembled from four round rods offers an appropriate effect. It was created a short time later and is also still in the current collection. It was added to the platinum-gold collection developed in 1998. In this case, four round rods, two thicker and two thinner ones, are connected in a way that creates a square that surrounds the finger. The somewhat longer, thicker round rods are positioned on top of the hand or on the palm side of the hand, and the thinner ones, inset a bit, connect the two visually operative elements. A technical specialty: the elements of this ring are not welded together but are connected with each other through press fits; only the platinum and gold versions need to be welded.

A short time later, still during the eighties, a group of rings was created that are worn exclusively on the pinky. They represent a functional specialty. Either three-quarter-

Two rings for the pinky. The three-quarter-round discs surround the small finger on the outside. Gold, stainless steel.

Ring open on the top with interrupted diameter of the band. They can be worn in pairs. Gold, stainless steel.

round semicircular discs or right angles formed from round rods are closed by semicircular arches and decorate the outside of the small finger. Whereas the other rings are designed and produced mainly with basic stereometric forms, the right angle and circular arch here take on the function that determines the look. They are interesting due to their constructive specialties, but they still remind one of intellectualized drawing board solutions whose practicality may indeed be questioned.

Finally, around 1990, a ring was designed that could be placed alongside the solution with the four round rods with respect to its solitary importance. We're talking about the band that is open at the top, whose one end cantilevers from the band. The massive band with a rectangular cross-section shows a clear incision after about three quarters of the way around its course, a separation joint reaching to half of the band's width. And right at this joint, the course of the band is not only interrupted, but broken – an almost dramatic event given the thickness of band. This ring can be worn in pairs since its potential counterpart fits precisely into the irregular course of the band, and it too was added to the new platinum-gold collection.

The oldest model, with the coat of arms look, is a constructive but logical solution; the square ring made of the four round rods is intelligent but may seem a bit strange as a ring, although when worn it will absolutely convince the wearer and admirer; but the pinky rings are missing the kind of self-understood manner of the one solution and the power of conviction of the others.

The Band Rings

In the early nineties, Dau developed the first group of band rings which in the ensuing years would become the dominant group of rings with respect to the number of models and demand.

As C.-F. Dau understands them, band rings are seamless bands with parallel outer edges that enclose the finger. At first glance they appear to be made of differently colored metal strips all lying on the same plane, or there are differently colored – also differently shaped, on occasion – bands resting on steel loops. And there are rings that are structured by fine, mostly vertical, recessed lines. The models are well proportioned and rather voluminous. They have a discreet sculptural structure. The wire loops or flat bands that are mounted onto the primary rings show only minimal, even if very consciously set, differences in levels. The sunken indentations have an effect that is more graphic than plastic: they separate the bands that make up the rings and ensure that the color of each material has its appropriate effect. All of the divisions – of color, shape, and graphics – follow the circular movement of the ring. They are inherent to the structure and create the horizontal pattern of the bands. The band rings are bipartite in their construction and in their possible effects. The oldest models are unchangeable. They are produced and sold in forms that were first concretely determined and then developed into a series. However, there is another group with an especially large number of variants: the variable rings. These are also band rings with parallel edges. They can be assembled from

Six band rings. The colored bands are pushed onto steel base volumes and are not removable. Gold, silver, stainless steel.

One of the band rings in three color versions.

Right page and following double page: The group of variable rings. The group is composed of seven basic forms that can be combined according to one's preference. The pictures show the possible coloration, from gray-white shades to combinations dominated by gold. The variable rings have become the best-selling ring group. Gold, silver, stainless steel, titanium.

seven different basic forms and generally consist of two or three, and sometimes even four individual loops. The loops lie loosely next to each other and are kept in place by the finger. They can be rearranged, their positions can be switched, and, provided the same ring size is used, they can be combined with other loops. The outer loops often gently slope down to the finger so that they can be worn comfortably and do not leave pressure marks on the finger. The wide variability of this group comes from the interchangeability of the individual loops and from the fact that all basic forms are produced in steel, silver, titanium, and gold. This great variability is probably not taken advantage of very often, but it is important for the assembly of the ring at the time of purchase.

Both types of band rings, the variable ones and the non-variable ones, are generally created from rod elements and machine-turned parts. At the Dau workshop, these are acquired as blanks and are refined, modified, and then either mechanically assembled or handmade to order. The mechanical assembly of the non-variable models requires special dimensional tolerances of the individual parts: They must fit together exactly. After assembly, the parts normally cannot be separated without being destroyed. In the case of the variable rings, the fittings are limited to the inner diameter, the ring width, and the surface alignment. Obviously, the width and thickness of the loops also must be kept within narrow tolerances since they constitute the unvarying appearance of the various patterns.

Over the years, the band rings and, in particular, the variable models, have proven to be the best-selling group. They are produced in considerable numbers and are often worn as wedding rings.

Three of seven individual rings that serve as the basic elements for the variable rings.

Following double page: A composition of the rings mentioned in platinum and gold.

Beautiful Serial Jewelry is Not Just a Design Task

The method of working and representation cultivated in the Dau workshop is far removed from handcrafted production. But, typically, it is also just as far removed from normal industrial production practices. Contrary to handcrafted production, and by purchasing all of the elements, attention is given uncompromisingly to the performance of modern machinery, precision and the always identical appearance of the machine-made products. Contrary to industrial production, however, where a quasi-craftsman's formal language is still being cultivated for large components, although all parts, and even whole pieces of jewelry, are produced with machines or castings, the machine-made production of the elements is not disguised here but made visible and emphasized. This work method, despite its straight-forwardness, has its consequences: machine-made parts can be purchased at a favorable price only if certain lots are produced with the help of the appropriate machines. If we additionally consider that for the production of several models not only several elements but elements made of different materials, among them precious metals, are needed, it becomes obvious that this method of working and production can be realized only on the basis of a relatively large and expensive inventory. The necessary size

of the inventory for these parts and the average, hardly calculable sale of the individual models highlight the problem inherent in this method and require a careful and clever organization of the enterprise and the development of the collection.

View into the generously arranged workshop of the goldsmith's business. Behind the workbench is a view of the side room, which also makes sense with respect to safety.

Between Unique Specimen and Mass Production

The comprehensive meaning of the jewelry from C.-F. Dau's workshop becomes clear if his collections are compared with the oversized, often overlapping offers by the industrial large producers, on one hand, and the handcrafted jewelry oriented towards the unique specimen on the other. With respect to clarity of the design, certainty and consequence, his work can unquestionably be compared with the intensity of the effect of handcrafted artistic jewelry. With respect to the technical conditions, Dau's solutions live up to the best serial products of industrial producers; in many cases, they actually outperform them. What distinguishes his work from the unique artistic specimen is the seemingly depersonalized factuality of the solutions as well as the design of the individual parts, freed from coincidence and emotion. Certainly, this design style follows the ambition of the designer; however, it mainly follows with a rather obligated necessity from the serial character of this jewelry. And it doesn't mind being reproduced – this is its destiny; it doesn't offer any manufactured pleasing emotions. Contrary to typical serial jewelry, Dau's works can be seen as textbook examples for proper manufacturing and material use. What looks like gold is indeed gold, and

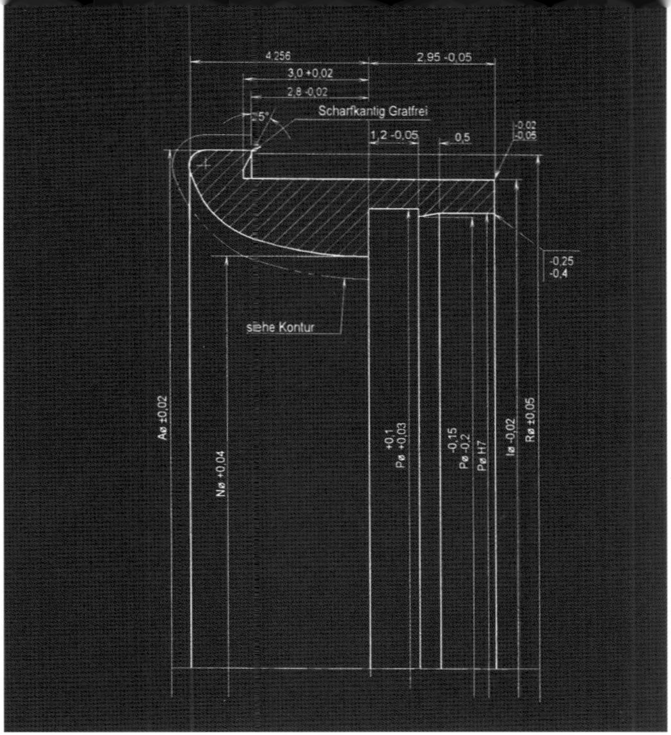

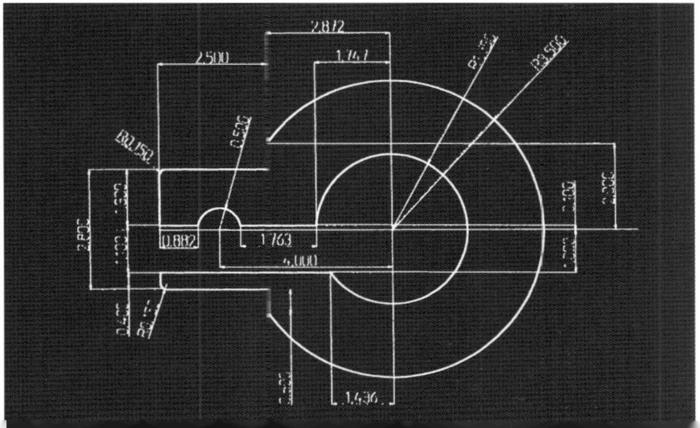

What design sketches are for the craftsman, technical drawings are for Dau's workshop. In them, the elements to be developed for serial production are described in detail, and they determine the communication with the producers of the elements.

View of a parts inventory: the blanks of a basic form of the variable rings, organized by ring size and material, can be clearly seen.

machine-turned parts are exactly what they seem. It is the lack of any imitation, the honesty in the design and technology, that distinguishes these pieces from all other compressed, stamped and cast pieces. And it is the emotional straight-forwardness that makes them sympathetic.

Of course, we cannot neglect the fact that the success of the collection is based on the lucky coincidence of a personal disposition and a specific taste for the times. In other words: the eighties and the goldsmith industry of that era were definitely primed for a designer like C.-F. Dau, and they were prepared to make his plans successful. But it is to his credit that he recognized this condition and used it for his designs with an increasing awareness.

Three stickpins from the latest platinum collection. The minimalist reserve of the design and, simultaneously, the extraordinary certainty in the formulation can hardly be surpassed.

Address:
Carl-F. Dau Dipl.-Des.
Goldschmiedemeister
Hohentwielsteig 10
D-14163 Berlin
Tel (030) 80 99 55 - 0
Fax (030) 80 99 55 44

1 Worringer, W.: *Abstraktion und Einfühlung. Ein Beitrag zur Stilpsychologie.* Dresden 1996.
2 Dau, C.-F.: Katalog 1, foreword.
3 See also, among others, Friedrich Knupper. *Schmuck und Objekte.* Edited by G. Bott. Nürnberg 1990.
4 Kállai, E.: „Serienkunst". In: *Das neue Frankfurt* 11/1930. Quoted from *Neues Bauen Neues Gestalten*. Dresden 1984.

C.-F. Dau on the roof terrace of the workshop and studio building at Hohentwielsteig.

Photo credits: Matthias Hoffmann, Düsseldorf.

Service
If you would like to receive our catalog, please contact us:

Verlag form. Books & Magazines on design issues.

Telephone +49 (0) 69 94 33 25-0
Facsimile +49 (0) 69 94 33 25-25
e-Mail: form@form.de
http://www.form.de

Impressum

Translated into English by:
Katja Steiner and Bruce Almberg

Graphic Design:
Shaun Wildsmith,
CBCD Cool Blue Corporate Design GmbH,
Frankfurt am Main

Lithography:
Hans Altenkirch
Mediaproduktion GmbH,
Nieder-Olm

Print:
Graspo,
Zlin, Tschechien

Die Deutsche Bibliothek –
CIP-Einheitsaufnahme
The Rings by C.-F.Dau / Hans-Ulrich Lehmann.
[Transl. into Engl. by Katja Steiner and Bruce Almberg]. – Frankfurt am Main : Verl. Form, 1999 (Design Classics)
Dt. Ausg. u.d.T.: Die Ringe von C.-F. Dau

ISBN 3-931317-69-2